Standards Success

Common Core State Standards Companion
for use with *Saxon Math Intermediate 4*

Copyright © by HMH Supplemental Publishers Inc.

All rights reserved. No part of this work may be reproduced or transmitted in any form or by any means, electronic or mechanical, including photocopying or recording, or by any information storage or retrieval system, without the prior written permission of the copyright owner unless such copying is expressly permitted by federal copyright law.

Permission is hereby granted to individuals using the corresponding student's textbook or kit as the major vehicle for regular classroom instruction to photocopy Lesson Extension Activities, Activity Masters, and Extension Tests from this publication in classroom quantities for instructional use and not for resale. Requests for information on other matters regarding duplication of this work should be addressed to Houghton Mifflin Harcourt Publishing Company, Attn: Contracts, Copyrights, and Licensing, 9400 South Park Center Loop, Orlando, Florida 32819.

Common Core State Standards © Copyright 2010. National Governors Association Center for Best Practices and Council of Chief State School Officers. All rights reserved.

This product is not sponsored or endorsed by the Common Core State Standards Initiative of the National Governors Association Center for Best Practices and the Council of Chief State School Officers.

Printed in the U.S.A.

ISBN 978-0-547-62815-8

6 7 8 9 10   1186 21 20 19 18 17 16

4500604150   A B C D E F G

If you have received these materials as examination copies free of charge, Houghton Mifflin Harcourt Publishing Company retains title to the materials and they may not be resold. Resale of examination copies is strictly prohibited.

Possession of this publication in print format does not entitle users to convert this publication, or any portion of it, into electronic format.

Table of Contents

Instructions for Using This Book

Overview i

***Saxon Math Intermediate 4* Table of Contents**
with Common Core State Standards References 1

Correlation to the Common Core State Standards 13

Lesson Extension Activities

Lesson 40 Extension Activity 1 19
- Activity Master 1

Investigation 4 Extension Activity 2 21
- Activity Master 2

Lesson 49 Extension Activity 3 23
- Activity Master 3

Lesson 81 Extension Activity 4 25
- Activity Master 4

Lesson 92 Extension Activity 5 27
- Activity Master 5

Lesson 104 Extension Activity 6 29
- Activity Master 6

Lesson 107 Extension Activity 7 31
- Activity Master 7

Lesson 108 Extension Activity 8 33
- Activity Master 8

Lesson 114 Extension Activity 9 35
- Activity Master 9

Lesson 116 – Extension Activity 10 37
- Activity Master 10

Extension Tests

Extension Test 1 – For use with Cumulative Test 8

Extension Test 2 – For use with Cumulative Test 9

Extension Test 3 – For use with Cumulative Test 10

Extension Test 4 – For use with Cumulative Test 16

Extension Test 5 – For use with Cumulative Test 19

Extension Test 6 – For use with Cumulative Test 20

Extension Test 7 – For use with Cumulative Test 21

Extension Test 8 – For use with Cumulative Test 22

Extension Test 9 – For use with Cumulative Test 23

Extension Test 10 – For use with Cumulative Test 23

Lesson Extension Activity Answers and Activity Master Answers 49

Extension Test Answers 50

Saxon Math *Intermediate 4*   © HMH Supplemental Publishers Inc.

Instructions for Using This Book

Educators who use Saxon Math know that the programs help students become competent and confident learners. Because of the incremental nature of the program, some lessons provide foundational instruction necessary for developing more advanced skills used in later lessons. The Power Up, Problem Solving, and Written Practice sections of each lesson provide important review and practice needed for mastery. For those reasons, it is essential to teach all the lessons in the correct order and to include all parts of the lesson in the daily instruction.

The program Table of Contents included in this book shows references to the primary Common Core State Standards domain and cluster or Mathematical Practice addressed by each lesson and investigation. The Lesson Extension Activities provided in this book will help reinforce that knowledge. Each of these activities was developed to spring from the instruction of the designated lesson or investigation.

It is recommended that you review the Table of Contents to understand where the extension lessons are to be integrated into the program. Then place a reminder in the Teacher's Manual, such as a colored flag or sticker, on the lesson or investigation with which the extension should be presented. Before the day the extension is to be taught, photocopy both the Lesson Extension Activity and Activity Master on the back (one copy of each per student). The pages are perforated to make removal and copying easier. At this time, also check the extension activity for any materials that are required and be prepared for each student to have the necessary items. The problems on the Activity Masters may be solved directly following the Lesson Extension Activity, or may be used as additional practice at a later time.

Extension Tests are provided to ensure that all Common Core objectives are evaluated. These multiple-choice assessments should be given with specified Cumulative Tests, as noted in the Table of Contents. Before test day, photocopy one copy of the Extension Test for each student's use.

Best wishes for a successful school year!

Saxon Math Intermediate 4 Standards Success
Overview

Common Core State Standards and the Saxon Math Pedagogy

The Saxon Math philosophy stresses that incremental and integrated instruction, with the opportunity to practice and internalize concepts, leads to successful mathematics understanding. This pedagogy aligns with the requirements of the Common Core State Standards, which emphasize that, in each grade, students will be instructed to mastery in specified math concepts that serve as a basis for future learning. For example, in Grade 4 students develop fluency in multi-digit mathematical operations and an understanding of fractions that can be carried forward to succeeding grades. Having established this solid foundation, the students will have the necessary tools (speed, accuracy, and confidence in their ability) to tackle increasingly complex problem solving.

The requisites featured in the Mathematical Practices are incorporated throughout the Saxon lessons and activities. For example, students are asked to share ideas and to think critically, to look for patterns, and to make connections in mathematical reasoning.

What *Saxon Math Intermediate 4 Standards Success* Provides

Saxon Math Intermediate 4 Standards Success is a companion to *Saxon Math Intermediate 4*. The first section, the Table of Contents, lists the Common Core focus of each lesson. The second section, Correlation of *Saxon Math Intermediate 4* to the Common Core State Standards for Mathematics Grade 4, demonstrates the depth of coverage provided by the *Saxon Math Intermediate 4* program. The remaining sections, Lesson Extension Activities and Extension Tests, provide additional reinforcement for selected Common Core standards.

Saxon Math Intermediate 4 Table of Contents

The Intermediate 4 Table of Contents lists the primary Common Core domain and cluster addressed in the New Concept of each lesson and that section's Investigation. Some lessons focus on a Mathematical Practice, such as a problem-solving technique. The primary Common Core State Standards focuses in the Power Up and Problem Solving activities of the ten lessons are listed on a chart at the bottom of each page of the Table of Contents.

Correlation of *Saxon Math Intermediate 4* to the Common Core State Standards for Mathematics Grade 4

The correlation lists the specific *Saxon Math Intermediate 4* components addressing each standard. This correlation is divided into three sections: Power Up (including Power Up and Problem Solving), Lessons (including New Concepts, Investigations, and Written Practices), and Other (including Calculator Activities, Performance Tasks, and Test Day Activities).

Lesson Extension Activities and *Extension Tests*

Lesson Extension Activities (with Activity Masters on the back) and Extension Tests are listed in the Table of Contents where they are intended to be used. These additional activities further address and reinforce the Common Core standards. Lesson Extension Activities, Activity Masters, and Extension Tests begin on page 19 of this book.

Saxon Math Intermediate 4 © HMH Supplemental Publishers Inc. i

Domains, Clusters, and Mathematical Practices for Grade 4

The Common Core State Standards are separated into domains, which are divided into clusters.

Grade 4 Domains and Clusters

Large groups of connected standards are referred to as domains. In Grade 4 there are five domains. Groups of related standards within a domain are referred to as clusters.

4.OA–Operations and Algebraic Thinking
- 1st cluster: Use the four operations with whole numbers to solve problems.
- 2nd cluster: Gain familiarity with factors and multiples.
- 3rd cluster: Generate and analyze patterns.

4.NBT–Number and Operations in Base Ten
- 1st cluster: Generalize place value understanding for multi-digit whole numbers.
- 2nd cluster: Use place value understanding and properties of operations to perform multi-digit arithmetic.

4.NF–Number and Operations—Fractions
- 1st cluster: Extend understanding of fraction equivalence and ordering.
- 2nd cluster: Build fractions from unit fractions by applying and extending previous understandings of operations on whole numbers.
- 3rd cluster: Understand decimal notation for fractions, and compare decimal fractions.

4.MD–Measurement and Data
- 1st cluster: Solve problems involving measurement and conversion of measurements from a larger unit to a smaller unit.
- 2nd cluster: Represent and interpret data.
- 3rd cluster: Geometric measurement: understand concepts of angle and measure angles.

4.G–Geometry
- 1st cluster: Draw and identify lines and angles, and classify shapes by properties of their lines and angles.

Mathematical Practices

The Standards for Mathematical Practice list the following essential competencies that students will develop throughout their mathematical education.

CC.K–12.MP.1 Make sense of problems and persevere in solving them.
CC.K–12.MP.2 Reason abstractly and quantitatively.
CC.K–12.MP.3 Construct viable arguments and critique the reasoning of others.
CC.K–12.MP.4 Model with mathematics.
CC.K–12.MP.5 Use appropriate tools strategically.
CC.K–12.MP.6 Attend to precision.
CC.K–12.MP.7 Look for and make use of structure.
CC.K–12.MP.8 Look for and express regularity in repeated reasoning.

For the full text of the Common Core State Standards and a comprehensive correlation, including Mathematical Practices, see the Correlation of *Saxon Math Intermediate 4* to the Common Core State Standards for Mathematics Grade 4 on pages 13–18.

Saxon Math Intermediate 4
TABLE OF CONTENTS

Lesson		Common Core State Standards Focus of Lesson
	Problem Solving Overview	CC.K–12.MP.1
1	• Review of Addition	CC.4.NBT (2nd cluster)
2	• Missing Addends	CC.4.OA (1st cluster)
3	• Sequences • Digits	CC.4.OA (3rd cluster)
4	• Place Value	CC.4.NBT (1st cluster)
5	• Ordinal Numbers • Months of the Year	CC.K–12.MP.2
6	• Review of Subtraction	CC.4.NBT (2nd cluster)
7	• Writing Numbers Through 999	CC.4.NBT (1st cluster)
8	• Adding Money	CC.4.NBT (2nd cluster)
9	• Adding with Regrouping	CC.4.NBT (2nd cluster)
10	• Even and Odd Numbers	CC.K–12.MP.4
Inv. 1	• Number Lines	CC.4.OA (3rd cluster)
	Cumulative Assessment	

The following table shows a CCSS* focus of the Power Up (PU) and the Problem Solving (PS) activities, which appear at the beginning of each lesson.

CCSS* Reference	1	2	3	4	5	6	7	8	9	10
CC.K–12.MP.1	PS	PS	PS	PS	PS	PS	PS	PS	PS	PS
CC.K–12.MP.4			PS	PS	PS				PS	PS
CC.4.OA (2nd cluster)										PU
CC.4.OA (3rd cluster)								PS		

*Common Core State Standards (CCSS)

Correlation references are read as follows: CC indicates Common Core, the number following is the grade, the letters indicate the domain, and the cluster indicates the particular group of related standards. Mathematical Practices are described in the same way for all grades K–12.

MP Mathematical Practices   **OA** Operations and Algebraic Thinking   **NBT** Number and Operations in Base Ten
NF Number and Operations—Fractions   **MD** Measurement and Data   **G** Geometry

Lesson		CCSS Focus of Lesson
11	• Addition Word Problems with Missing Addends	CC.4.OA (1st cluster)
12	• Missing Numbers in Subtraction	CC.4.NBT (2nd cluster)
13	• Adding Three-Digit Numbers	CC.4.NBT (2nd cluster)
14	• Subtracting Two-Digit and Three-Digit Numbers • Missing Two-Digit Addends	CC.4.NBT (2nd cluster)
15	• Subtracting Two-Digit Numbers with Regrouping	CC.4.NBT (2nd cluster)
	Cumulative Assessment	
16	• Expanded Form • More on Missing Numbers in Subtraction	CC.4.NBT (1st cluster) CC.4.NBT (2nd cluster)
17	• Adding Columns of Numbers with Regrouping	CC.4.NBT (2nd cluster)
18	• Temperature	CC.K–12.MP.5
19	• Elapsed Time Problems	CC.4.MD (1st cluster)
20	• Rounding	CC.4.NBT (1st cluster)
Inv. 2	• Units of Length and Perimeter	CC.4.MD (1st cluster)
	Cumulative Assessment	

The following table shows a CCSS focus of the Power Up (PU) and the Problem Solving (PS) activities, which appear at the beginning of each lesson.

CCSS Reference	11	12	13	14	15	16	17	18	19	20
CC.K–12.MP.1	PS	PS	PS	PS	PS	PS	PS	PS	PS	PS
CC.K–12.MP.3										PS
CC.K–12.MP.4						PS				
CC.K–12.MP.7										PS
CC.4.OA (2nd cluster)			PU	PU		PU		PU	PU	
CC.4.OA (3rd cluster)	PS	PS								
CC.4.NBT (2nd cluster)		PU	PU	PU	PU	PU	PU	PU	PU	PU

Lesson		CCSS Focus of Lesson
21	• Triangles, Rectangles, Squares, and Circles	CC.4.G (1st cluster)
22	• Naming Fractions • Adding Dollars and Cents	CC.4.NF (1st cluster) CC.K–12.MP.4
23	• Lines, Segments, Rays, and Angles	CC.4.G (1st cluster)
24	• Inverse Operations	CC.4.NBT (2nd cluster)
25	• Subtraction Word Problems	CC.4.NBT (2nd cluster)
	Cumulative Assessment	
26	• Drawing Pictures of Fractions	CC.K–12.MP.4
27	• Multiplication as Repeated Addition • More Elapsed Time Problems	CC.K–12.MP.8 CC.4.MD (1st cluster)
28	• Multiplication Table	CC.4.NBT (2nd cluster)
29	• Multiplication Facts: 0s, 1s, 2s, 5s	CC.4.NBT (2nd cluster)
30	• Subtracting Three-Digit Numbers with Regrouping	CC.4.NBT (2nd cluster)
Inv. 3	• Multiplication Patterns • Area • Squares and Square Roots	CC.4.NBT (2nd cluster)
	Cumulative Assessment	

The following table shows a CCSS focus of the Power Up (PU) and the Problem Solving (PS) activities, which appear at the beginning of each lesson.

CCSS Reference	21	22	23	24	25	26	27	28	29	30
CC.K–12.MP.1	PS	PS	PS	PS	PS	PS	PS	PS	PS	PS
CC.K–12.MP.7					PS					
CC.K–12.MP.8								PS		
CC.4.OA (1st cluster)								PS		PS
CC.4.OA (2nd cluster)	PU		PU							
CC.4.OA (3rd cluster)		PS		PS	PS					PS
CC.4.NBT (2nd cluster)	PU	PU	PU		PU	PU	PU	PU	PU	PU
CC.4.MD (1st cluster)		PU	PU	PU	PU			PU	PU	PU
CC.4.G (1st cluster)							PS			

MP Mathematical Practices   **OA** Operations and Algebraic Thinking   **NBT** Number and Operations in Base Ten
NF Number and Operations—Fractions   **MD** Measurement and Data   **G** Geometry

Saxon Math Intermediate 4   © HMH Supplemental Publishers Inc.

Lesson		CCSS Focus of Lesson
31	• Word Problems About Comparing	CC.4.OA (1st cluster)
32	• Multiplication Facts: 9s, 10s, 11s, 12s	CC.4.NBT (2nd cluster)
33	• Writing Numbers Through Hundred Thousands	CC.4.NBT (1st cluster)
34	• Writing Numbers Through Hundred Millions	CC.4.NBT (1st cluster)
35	• Naming Mixed Numbers and Money	CC.4.NF (3rd cluster)
	Cumulative Assessment	
36	• Fractions of a Dollar	CC.4.NF (1st cluster)
37	• Reading Fractions and Mixed Numbers from a Number Line	CC.K–12.MP.6
38	• Multiplication Facts (Memory Group)	CC.4.NBT (2nd cluster)
39	• Reading an Inch Scale to the Nearest Fourth	CC.K–12.MP.6
40	• Capacity *Lesson Extension Activity 1 (p 19):* • Using Diagrams to Solve Problems	CC.4.MD (1st cluster)
Inv. 4A	• Tenths and Hundredths	CC.4.NF (3rd cluster)
Inv. 4B	• Relating Fractions and Decimals *Lesson Extension Activity 2 (p 21):* • Adding Fractions with Denominators 10 and 100	CC.4.NF (3rd cluster)
	Cumulative Assessment	

The following table shows a CCSS focus within the Power Up (PU) and Problem Solving (PS) activities, which appear at the beginning of each lesson.

CCSS Reference	31	32	33	34	35	36	37	38	39	40
CC.K–12.MP.1	PS	PS	PS	PS	PS	PS	PS	PS	PS	PS
CC.K–12.MP.4		PS		PS		PS			PS	
CC.K–12.MP.8	PS									
CC.4.OA (3rd cluster)	PS		PS		PS					
CC.4.NBT (1st cluster)		PU	PU	PU						
CC.4.NBT (2nd cluster)		PU			PU	PU		PS		
CC.4.MD (1st cluster)	PU	PU	PU		PU	PU			PU	PU

Lesson		CCSS Focus of Lesson
41	• Subtracting Across Zero • Missing Factors	CC.4.NBT (2nd cluster) CC.4.OA (1st cluster)
42	• Rounding Numbers to Estimate	CC.4.NBT (1st cluster)
43	• Adding and Subtracting Decimal Numbers, Part 1	CC.4.MD (1st cluster)
44	• Multiplying Two-Digit Numbers, Part 1	CC.4.NBT (2nd cluster)
45	• Parentheses and the Associative Property • Naming Lines and Segments	CC.4.NBT (2nd cluster) CC.4.G (1st cluster)
	Cumulative Assessment *Extension Test 1*	
46	• Relating Multiplication and Division, Part 1	CC.4.NBT (2nd cluster)
47	• Relating Multiplication and Division, Part 2	CC.4.NBT (2nd cluster)
48	• Multiplying Two-Digit Numbers, Part 2	CC.4.NBT (2nd cluster)
49	• Word Problems About Equal Groups, Part 1 Lesson Extension Activity 3 (p 23): • Solving Comparison Problems	CC.4.NBT (2nd cluster) CC.4.OA (1st cluster)
50	• Adding and Subtracting Decimal Numbers, Part 2	CC.K–12.MP.6
Inv. 5	• Percents	CC.4.NF (3rd cluster)
	Cumulative Assessment *Extension Test 2*	

The following table shows a CCSS focus of the Power Up (PU) and the Problem Solving (PS) activities, which appear at the beginning of each lesson.

CCSS Reference	41	42	43	44	45	46	47	48	49	50
CC.K–12.MP.1	PS	PS	PS	PS	PS	PS	PS	PS	PS	PS
CC.K–12.MP.2								PS		
CC.K–12.MP.5				PS						
CC.K–12.MP.6		PU/PS	PU		PU	PU	PU	PU	PU	PU
CC.K–12.MP.7										PS
CC.4.OA (3rd cluster)	PS									
CC.4.NBT (2nd cluster)									PU	PU
CC.4.MD (1st cluster)	PU	PU/PS	PU	PU	PU	PU				PU

MP Mathematical Practices   **OA** Operations and Algebraic Thinking   **NBT** Number and Operations in Base Ten
NF Number and Operations—Fractions   **MD** Measurement and Data   **G** Geometry

Saxon Math Intermediate 4   © HMH Supplemental Publishers Inc.   **5**

Lesson		CCSS Focus of Lesson
51	• Adding Numbers with More Than Three Digits • Checking One-Digit Division	CC.4.NBT (2nd cluster)
52	• Subtracting Numbers with More Than Three Digits • Word Problems with Equal Groups, Part 2	CC.4.NBT (2nd cluster)
53	• One-Digit Division with a Remainder	CC.4.NBT (2nd cluster)
54	• The Calendar • Rounding Numbers to the Nearest Thousand	CC.4.NBT (1st cluster)
55	• Prime and Composite Numbers	CC.4.OA (2nd cluster)
	Cumulative Assessment *Extension Test 3*	
56	• Using Models and Pictures to Compare Fractions	CC.4.NF (1st cluster)
57	• Rate Word Problems	CC.4.OA (1st cluster)
58	• Multiplying Three-Digit Numbers	CC.4.NBT (2nd cluster)
59	• Estimating Arithmetic Answers	CC.4.NBT (2nd cluster)
60	• Rate Problems with a Given Total	CC.4.OA (1st cluster)
Inv. 6	• Displaying Data Using Graphs	CC.4.MD (2nd cluster)
	Cumulative Assessment	

The following table shows a CCSS focus of the Power Up (PU) and the Problem Solving (PS) activities, which appear at the beginning of each lesson.

CCSS Reference	51	52	53	54	55	56	57	58	59	60
CC.K–12.MP.1	PS	PS	PS	PS	PS	PS	PS	PS	PS	PS
CC.K–12.MP.4								PS		
CC.K–12.MP.6	PU	PU			PU	PU		PU	PU	
CC.K–12.MP.7			PU	PU	PS		PS			
CC.4.MD (1st cluster)	PS				PU	PS	PU	PU		

Lesson		CCSS Focus of Lesson
61	• Remaining Fraction • Two-Step Equations	CC.4.NF (2nd cluster) CC.4.OA (1st cluster)
62	• Multiplying Three or More Factors • Exponents	CC.4.NBT (2nd cluster) CC.K–12.MP.7
63	• Polygons	CC.4.G (1st cluster)
64	• Division with Two-Digit Answers, Part 1	CC.4.NBT (2nd cluster)
65	• Division with Two-Digit Answers, Part 2	CC.4.NBT (2nd cluster)
	Cumulative Assessment	
66	• Similar and Congruent Figures	CC.K–12.MP.8
67	• Multiplying by Multiples of 10	CC.4.NBT (2nd cluster)
68	• Division with Two-Digit Answers and a Remainder	CC.4.NBT (2nd cluster)
69	• Millimeters	CC.4.MD (1st cluster)
70	• Word Problems About a Fraction of a Group	CC.4.NF (2nd cluster)
Inv. 7	• Collecting Data with Surveys	CC.4.MD (2nd cluster)
	Cumulative Assessment	

The following table shows a CCSS focus of the Power Up (PU) and the Problem Solving (PS) activities, which appear at the beginning of each lesson.

CCSS Reference	61	62	63	64	65	66	67	68	69	70
CC.K–12.MP.1	PS	PS	PS	PS	PS	PS	PS	PS	PS	PS
CC.K–12.MP.2			PU		PS		PS		PS	
CC.K–12.MP.4		PU						PU		PS
CC.K–12.MP.6			PU	PS		PU				
CC.4.OA (3rd cluster)		PS								
CC.4.NBT (1st cluster)	PU	PU				PU		PU	PU	
CC.4.MD (1st cluster)		PU					PS	PU		PU

MP Mathematical Practices   **OA** Operations and Algebraic Thinking   **NBT** Number and Operations in Base Ten
NF Number and Operations—Fractions   **MD** Measurement and Data   **G** Geometry

Saxon Math Intermediate 4   © HMH Supplemental Publishers Inc.   **7**

Lesson		CCSS Focus of Lesson
71	• Division Answers Ending with Zero	CC.4.NBT (2nd cluster)
72	• Finding Information to Solve Problems	CC.K–12.MP.1
73	• Geometric Transformations	CC.K–12.MP.4
74	• Fraction of a Set	CC.4.NF (2nd cluster)
75	• Measuring Turns	CC.4.MD (3rd cluster)
	Cumulative Assessment	
76	• Division with Three-Digit Answers	CC.4.NBT (2nd cluster)
77	• Mass and Weight	CC.4.MD (1st cluster)
78	• Classifying Triangles	CC.4.G (1st cluster)
79	• Symmetry	CC.4.G (1st cluster)
80	• Division with Zeros in Three-Digit Answers	CC.4.NBT (2nd cluster)
Inv. 8	• Analyzing and Graphing Relationships	CC.4.MD (2nd cluster)
	Cumulative Assessment	

The following table shows a CCSS focus of the Power Up (PU) and the Problem Solving (PS) activities, which appear at the beginning of each lesson.

CCSS Reference	71	72	73	74	75	76	77	78	79	80
CC.K–12.MP.1	PS	PS	PS	PS	PS	PS	PS	PS	PS	PS
CC.K–12.MP.2	PS		PS					PS		
CC.K–12.MP.5							PS			
CC.4.OA (1st cluster)	PS						PS			PS
CC.4.NBT (1st cluster)		PU	PU	PU		PU				
CC.4.MD (1st cluster)			PS	PU	PU		PU	PU/PS		

8   © HMH Supplemental Publishers Inc.   *Saxon Math* Intermediate 4

Lesson		CCSS Focus of Lesson
81	• Angle Measures Lesson Extension Activity 4 (p 25): • Measuring and Drawing Angles	CC.4.G (1st cluster) CC.4.MD (3rd cluster)
82	• Tessellations	CC.K–12.MP.4
83	• Sales Tax	CC.4.MD (1st cluster)
84	• Decimal Numbers to Thousandths	CC.4.NF (3rd cluster)
85	• Multiplying by 10, by 100, and by 1000	CC.4.NBT (2nd cluster)
	Cumulative Assessment *Extension Test 4*	
86	• Multiplying Multiples of 10 and 100	CC.4.NBT (2nd cluster)
87	• Multiplying Two Two-Digit Numbers, Part 1	CC.4.NBT (2nd cluster)
88	• Remainders in Word Problems About Equal Groups	CC.4.OA (1st cluster)
89	• Mixed Numbers and Improper Fractions	CC.4.NF (2nd cluster)
90	• Multiplying Two Two-Digit Numbers, Part 2	CC.4.NBT (2nd cluster)
Inv. 9	• Investigating Fractions with Manipulatives	CC.4.NF (2nd cluster)
	Cumulative Assessment	

The following table shows a CCSS focus of the Power Up (PU) and the Problem Solving (PS) activities, which appear at the beginning of each lesson.

CCSS Reference	81	82	83	84	85	86	87	88	89	90
CC.K–12.MP.1	PS	PS	PS	PS	PS	PS	PS	PS	PS	PS
CC.K–12.MP.2							PS			
CC.K–12.MP.4		PS								
CC.K–12.MP.5		PS						PS		
CC.K–12.MP.6	PU		PU			PU		PU	PU	PS
CC.4.OA (1st cluster)			PS							
CC.4.NBT (1st cluster)	PU	PU	PU						PU	
CC.4.MD (1st cluster)	PU					PU	PU/PS	PU	PU	

MP Mathematical Practices   **OA** Operations and Algebraic Thinking   **NBT** Number and Operations in Base Ten
NF Number and Operations—Fractions   **MD** Measurement and Data   **G** Geometry

Saxon Math Intermediate 4   © HMH Supplemental Publishers Inc.

Lesson		CCSS Focus of Lesson
91	• Decimal Place Value	CC.4.NF (3rd cluster)
92	• Classifying Quadrilaterals	CC.4.G (1st cluster)
	Lesson Extension Activity 5 (p 27):	
	• Joining and Separating Angles	CC.4.MD (3rd cluster)
93	• Estimating Multiplication and Division Answers	CC.4.OA (1st cluster)
94	• Two-Step Word Problems	CC.4.OA (1st cluster)
95	• Two-Step Problems About a Fraction of a Group	CC.4.NF (2nd cluster)
	Cumulative Assessment	
96	• Average	CC.4.OA (1st cluster)
97	• Mean, Median, Range, and Mode	CC.K–12.MP.8
98	• Geometric Solids	CC.K–12.MP.8
99	• Constructing Prisms	CC.K–12.MP.4
100	• Constructing Pyramids	CC.K–12.MP.4
Inv. 10	• Probability	CC.K–12.MP.7
	Cumulative Assessment	
	Extension Test 5	

The following table shows a CCSS focus of the Power Up (PU) and the Problem Solving (PS) activities, which appear at the beginning of each lesson.

CCSS Reference	91	92	93	94	95	96	97	98	99	100
CC.K–12.MP.1	PS	PS	PS	PS	PS	PS	PS	PS	PS	PS
CC.K–12.MP.2					PS					
CC.K–12.MP.5	PS									
CC.K–12.MP.6										PS
CC.4.OA (1st cluster)									PS	
CC.4.OA (3rd cluster)		PS		PS						
CC.4.NBT (1st cluster)	PU				PU					
CC.4.MD (1st cluster)			PU							

Lesson		CCSS Focus of Lesson
101	• Tables and Schedules	CC.4.MD (1st cluster)
102	• Tenths and Hundredths on a Number Line	CC.4.MD (1st cluster)
103	• Fractions Equal to 1 and Fractions Equal to $\frac{1}{2}$	CC.4.NF (1st cluster)
104	• Changing Improper Fractions to Whole or Mixed Numbers Lesson Extension Activity 6 (p 29): • Multiplying a Fraction by a Whole Number	CC.4.NF (1st cluster) CC.4.NF (2nd cluster)
105	• Dividing by 10	CC.4.NBT (2nd cluster)
	Cumulative Assessment *Extension Test 6*	
106	• Evaluating Expressions	CC.K–12.MP.7
107	• Adding and Subtracting Fractions with Common Denominators Lesson Extension Activity 7 (p 31): • Making a Line Plot to Display a Data Set	CC.4.NF (2nd cluster) CC.4.MD (2nd cluster)
108	• Formulas • Distributive Property Lesson Extension Activity 8 (p 33): • Finding Unknown Measures	CC.4.MD (1st cluster)
109	• Equivalent Fractions	CC.4.NF (1st cluster)
110	• Dividing by Multiples of 10	CC.4.NBT (2nd cluster)
Inv. 11	• Volume	CC.K–12.MP.6
	Cumulative Assessment *Extension Test 7*	

The following table shows a CCSS focus of the Power Up (PU) and the Problem Solving (PS) activities, which appear at the beginning of each lesson.

CCSS Reference	101	102	103	104	105	106	107	108	109	110
CC.K–12.MP.1	PS	PS	PS	PS	PS	PS	PS	PS	PS	PS
CC.K–12.MP.3		PS								
CC.K–12.MP.6				PS						
CC.4.OA (1st cluster)	PS			PS						
CC.4.OA (3rd cluster)					PS			PS		
CC.4.NBT (1st cluster)	PU			PU			PU			
CC.4.NBT (2nd cluster)	PS									
CC.4.NF (2nd cluster)							PU	PU	PU	PU	PU
CC.4.MD (1st cluster)				PU						

MP Mathematical Practices   **OA** Operations and Algebraic Thinking   **NBT** Number and Operations in Base Ten
NF Number and Operations—Fractions   **MD** Measurement and Data   **G** Geometry

Saxon Math Intermediate 4   © HMH Supplemental Publishers Inc.

Lesson		CCSS Focus of Lesson
111	• Estimating Perimeter, Area, and Volume	CC.4.MD (1st cluster)
112	• Reducing Fractions	CC.4.NF (1st cluster)
113	• Multiplying a Three-Digit Number by a Two-Digit Number	CC.4.NBT (2nd cluster)
114	• Simplifying Fraction Answers Lesson Extension Activity 9 (p 35): • Adding and Subtracting Mixed Numbers	CC.4.NF (1st cluster) CC.4.NF (2nd cluster)
115	• Renaming Fractions	CC.4.NF (1st cluster)
	Cumulative Assessment *Extension Test 8*	
116	• Common Denominators Lesson Extension Activity 10 (p 37): • Comparing Fractions Using Common Denominators	CC.4.NF (2nd cluster) CC.4.NF (1st cluster)
117	• Rounding Whole Numbers Through Hundred Millions	CC.4.NBT (1st cluster)
118	• Dividing by Two-Digit Numbers	CC.4.NBT (2nd cluster)
119	• Adding and Subtracting Fractions with Different Denominators	CC.4.NF (2nd cluster)
120	• Adding and Subtracting Mixed Numbers with Different Denominators	CC.4.NF (2nd cluster)
Inv. 12	• Solving Balanced Equations	CC.K–12.MP.4
	Cumulative Assessment *Extension Tests 9 and 10*	

The following table shows a CCSS focus of the Power Up (PU) and the Problem Solving (PS) activities, which appear at the beginning of each lesson.

CCSS Reference	111	112	113	114	115	116	117	118	119	120
CC.K–12.MP.1	PS	PS	PS	PS	PS	PS	PS	PS	PS	PS
CC.4.OA (1st cluster)				PS						
CC.4.OA (3rd cluster)		PS	PS		PS		PS			PS
CC.4.NBT (1st cluster)					PU	PU				
CC.4.NBT (2nd cluster)				PS						
CC.4.NF (2nd cluster)	PU	PU				PU/PS				
CC.4.MD (1st cluster)				PU						

MP Mathematical Practices   **OA** Operations and Algebraic Thinking   **NBT** Number and Operations in Base Ten
NF Number and Operations—Fractions   **MD** Measurement and Data   **G** Geometry

© HMH Supplemental Publishers Inc.   ***Saxon Math*** *Intermediate 4*

Correlation of *Saxon Math Intermediate 4* to the Common Core State Standards for Mathematics Grade 4

Mathematical Practices – *These standards are covered throughout the program; the following are examples.*

1.	Make sense of problems and persevere in solving them.	**Power Up:** PS6, PS15, PS17, PS21, PS26, PS29, PS37, PS38, PS49, PS56, PS65, PS69, PS71, PS73, PS77, PS83, PS87, PS95, PS99, PS104, PS119 **Lessons:** Problem-Solving Overview (pp 1–6), L11, WP14, L25, WP28, L31, WP46, L49, L52, L57, WP67, L70, L72, Inv8, L88, WP88, L94, L95
2.	Reason abstractly and quantitatively.	**Power Up:** PS65, PS69, PS71, PS73, PS78, PS87, PS95 **Lessons:** L1, L11, Inv2, L25, L31, L32, WP46, L49, L52, L57, L58, L60, L70, WP77, WP78, WP84, L88, WP92, L94, L95, WP104, WP106 **Other:** PT3, TDA3
3.	Construct viable arguments and critique the reasoning of others.	**Power Up:** PS20, PS48, PS102 **Lessons:** L10, Inv1, L13, L15, WP15, L22, Inv3, WP67, L78, L83, L93 **Other:** PT5, TDA8, TDA9, TDA11
4.	Model with mathematics.	**Power Up:** PS3, PS4, PS5, PS9, PS10, PS16, PS32, PS34, PS36, PS39, PS58, PS70, PS82 **Lessons:** L11, L13, Inv2, WP25, Inv3, L38, L40, Inv4, L57, L60, Inv6, WP65, WP68, L70, WP70, Inv7, Inv8, WP84, WP92, L95, WP102, WP104, L108, Inv11
5.	Use appropriate tools strategically.	**Power Up:** PS44, PS77, PS82, PS88, PS91 **Lessons:** L18, Inv2, L21, L39, L51, L55, L69, Inv7, L77, L79, L80, L81, L101, L102, L103, L104 **Other:** CA2, CA22, TDA2, CA46, CA57, CA87, CA91, CA112
6.	Attend to precision.	**Power Up:** PS42, PS64, PS90, PS104 **Lessons:** Inv1, L18, L19, Inv2, Inv3, L39, L69, L75, WP79, WP84, WP96, L101, L102, WP102, WP108, Inv11
7.	Look for and make use of structure.	**Power Up:** PS20, PS25, PS50, PS55 **Lessons:** L9, Inv1, L13, L28, WP28, Inv3, L38, L45, WP46, L52, L62, L65, L87, L108, L110 **Other:** TDA4
8.	Look for and express regularity in repeated reasoning.	**Power Up:** PS28, PS31, PS100 **Lessons:** L20, L27, L29, Inv3, L42, L44, L55, L58, L59, L67, L85, L86, L90, L109, L112, L115, L116

Common Core State Standards	***Saxon Math Intermediate 4*** *Italic references indicate foundational.*
Operations and Algebraic Thinking 4.OA	
Use the four operations with whole numbers to solve problems.	
1. Interpret a multiplication equation as a comparison, e.g., interpret $35 = 5 \times 7$ as a statement that 35 is 5 times as many as 7 and 7 times as many as 5. Represent verbal statements of multiplicative comparisons as multiplication equations.	**Lessons:** L27, L28, L29, L38(TE), L46, L47, WP47–WP53, WP56–WP58, WP60–WP63 **Other:** LS46, LS47, LXA3, ET3

Key:
- **BT:** Benchmark Test
- **CA:** Calculator Activity
- **CT:** Cumulative Test
- **ECE:** End-of-Course Exam
- **ET:** Extension Test
- **Inv:** Investigation
- **L:** Lesson
- **LS:** Learning Station
- **LXA:** Lesson Extension Activity
- **PS:** Problem Solving
- **PT:** Performance Task
- **PU:** Power Up
- **TDA:** Test-Day Activity
- **WP:** Written Practice

Saxon Math Intermediate 4

COMMON CORE **Common Core State Standards**	*Saxon Math Intermediate 4* *Italic references indicate foundational.*
2. Multiply or divide to solve word problems involving multiplicative comparison, e.g., by using drawings and equations with a symbol for the unknown number to represent the problem, distinguishing multiplicative comparison from additive comparison.	**Power Up:** PS28, PS30, PS104 **Lessons:** *L41*, WP41–WP45, *L46*, *L47*, *L49*, WP49, WP51, *L52*, WP52, *L53*, WP54, WP57, *L60*, WP60, WP62, *L64*, WP64, WP66, WP67, *L68*, WP68–WP70, WP72–WP74, WP76–WP82, WP85, WP86, WP87, WP89, WP92–WP97, WP100, WP105, WP107, WP108, WP109 **Other:** LXA3, CT10, ET3, LS60, CT11, CT12, CT13, CT14, CT15, CT16, CT18, CT19, CT20, CT21, CT22, ECE
3. Solve multistep word problems posed with whole numbers and having whole-number answers using the four operations, including problems in which remainders must be interpreted. Represent these problems using equations with a letter standing for the unknown quantity. Assess the reasonableness of answers using mental computation and estimation strategies including rounding.	**Power Up:** PS71, PS77, PS80, PS83, PS99, PS101, PS104, PS114 **Lessons:** L59, L60, *L61*, WP61–WP63, *L64*, WP64, L65, WP65, WP66, WP68, WP69, WP71, WP73, WP75–WP78, L80, WP80–WP82, L83, WP83–WP87, L88, WP88–WP91, WP93, L94, WP94–WP106, WP108–WP116, WP118, WP120 **Other:** PT4, LS61, LS94, CT19, CT20

Gain familiarity with factors and multiples.

4. Find all factor pairs for a whole number in the range 1–100. Recognize that a whole number is a multiple of each of its factors. Determine whether a given whole number in the range 1–100 is a multiple of a given one-digit number. Determine whether a given whole number in the range 1–100 is prime or composite.	**Power Up:** PU10, PU13, PU14, PU16, PU18, PU19, PU21, PU23 **Lessons:** L55, WP55–WP64, WP67, WP69–WP74, WP78, WP79, WP84, WP89, WP111, WP112 **Other:** LS55, CT11, BT5, ECE

Generate and analyze patterns.

5. Generate a number or shape pattern that follows a given rule. Identify apparent features of the pattern that were not explicit in the rule itself.	**Power Up:** PS8, PS11, PS12, PS22, PS24, PS25, PS30, PS31, PS33, PS35, PS41, PS62, PS92, PS94, PS105, PS108, PS112, PS113, PS115, PS117, PS120 **Lessons:** L3, WP3–WP10, *Inv1*, WP11–WP19, WP21–WP25, WP27–WP29, Inv3, L32, WP32–WP36, L38, WP38, WP39, WP41, WP42, WP54 **Other:** LS3, CT1, CT2, CT3, BT1, CT5, CT6, TDA3, BT2, ECE

Number and Operations in Base Ten[1] 4.NBT

Generalize place value understanding for multi-digit whole numbers.

1. Recognize that in a multi-digit whole number, a digit in one place represents ten times what it represents in the place to its right.	**Lessons:** L4, WP4–WP7, WP9, WP12, L13, WP13–WP18, WP20–WP22, WP24–WP26, WP28, L33, WP33, L34, WP34–WP40, WP42, WP44, WP45, WP47–WP49, L50, WP50, L67, L85, WP85, L86, WP91, WP98, WP100, WP101, WP104, L105, WP105, WP107–WP109 **Other:** LS4, LS33, LS34, LS67, CT17, CT18, CT19, BT5, CT20, CT21, CT22, ECE
2. Read and write multi-digit whole numbers using base-ten numerals, number names, and expanded form. Compare two multi-digit numbers based on meanings of the digits in each place, using >, =, and < symbols to record the results of comparisons.	**Lessons:** L4, WP4–WP12, L13, WP13–WP15, L16, WP16–WP18, WP20–WP26, WP28–WP32, L33, WP33, L34, WP34–WP45, WP47–WP50 **Other:** LS4, CT1, CT2, TDA1, CT3, BT1, CT4, LS33, LS34, CT6, CT7, BT2, CT8, ECE

[1]Grade 4 expectations in this domain are limited to whole numbers less than or equal to 1,000,000.

COMMON CORE **Common Core State Standards**	*Saxon Math Intermediate 4* *Italic references indicate foundational.*
3. Use place value understanding to round multi-digit whole numbers to any place.	**Power Up:** PU32–PU34, PU61, PU62, PU66, PU68, PU69, PU72–PU74, PU76, PU81–PU83, PU89, PU91, PU95, PU101, PU104, PU107, PU115, PU116 **Lessons:** L20, WP20–WP30, WP32, WP34–WP38, L42, WP42–WP44, WP46, WP47, WP51–WP53, L54, WP54–WP56, WP66, WP69, WP102, WP104, WP106, WP107, WP110, WP113, L117, WP117–WP119 **Other:** CT4, CT5, CT7, PT4, BT2, LS42, CT8, CT9, LS54, CT10, CT11, TDA6, PT9, LS117

Use place value understanding and properties of operations to perform multi-digit arithmetic.

4. Fluently add and subtract multi-digit whole numbers using the standard algorithm.	**Power Up:** PU12–PU23, PU25–PU30, PU32, PU35, PU36, PS38, PU49, PU50, PS101 **Lessons:** L13, WP13, L14, WP14, L15, WP15, WP16, L17, WP17–WP21, WP23, WP24, L25, WP25–WP29, L30, WP30, L31, WP31–WP40, L41, WP41, WP42, L51, WP51, L52, WP52, WP54, WP56–WP58, L59, WP59, WP62, WP64, WP66, WP67, WP69–WP71, WP73–WP75, WP77, WP80–WP83, WP86, WP89, L94, WP96, WP103–WP106 **Other:** LS13, LS14, LS17, CT3, BT1, CT4, CT5, CT6, CT7, LS41, CT8, LS51, CT10, CT11, BT3, CT12, CT13, CT15, CT17, CT18, BT5, ECE
5. Multiply a whole number of up to four digits by a one-digit whole number, and multiply two two-digit numbers, using strategies based on place value and the properties of operations. Illustrate and explain the calculation by using equations, rectangular arrays, and/or area models.	**Lessons:** *L28, Inv3,* L42, WP42, WP43, L44, WP44, L45, WP45–WP47, *L48,* WP48–WP57, L58, WP58–WP61, L62, WP62–WP66, L67, WP68–WP75, WP77–WP80, WP82–WP86, L87, WP87–WP89, L90, WP90–WP92, WP94–WP107, L108, WP111, WP116, WP117, WP119 **Other:** LS44, LS45, CT9, CT10, LS58, CT11, BT3, CT12, LS67, CT13, CT14, BT4, LS87, LS90, CT17, CT18, CT19, BT5, CT20, CT21, CT22, ECE
6. Find whole-number quotients and remainders with up to four-digit dividends and one-digit divisors, using strategies based on place value, the properties of operations, and/or the relationship between multiplication and division. Illustrate and explain the calculation by using equations, rectangular arrays, and/or area models.	**Power Up:** PS114 **Lessons:** *L46, L47, L53,* L64, WP64, L65, WP65–WP67, L68, WP68–WP70, L71, WP71–WP75, L76, WP76–WP79, L80, WP80–WP113, WP115–WP120 **Other:** CT11, LS64, LS65, CT12, LS68, CT13, LS71, CT14, LS76, LS80, CT15, CT16, CT19, CT22, CT23, ECE

Number and Operations—Fractions[2] 4.NF

Extend understanding of fraction equivalence and ordering.

1. Explain why a fraction $\frac{a}{b}$ is equivalent to a fraction $\frac{(n \times a)}{(n \times b)}$ by using visual fraction models, with attention to how the number and size of the parts differ even though the two fractions themselves are the same size. Use this principle to recognize and generate equivalent fractions.	**Lessons:** Inv9, L103, L109, WP109, WP111, L112, WP112–WP114, L115, WP115, L116, WP116–WP118, L119, WP119, L120, WP120 **Other:** LS109, LS115, CT22, CT23, ECE

[2]Grade 4 expectations in this domain are limited to fractions with denominators 2, 3, 4, 5, 6, 8, 10, 12, and 100.

Key:	**BT:** Benchmark Test	**ET:** Extension Test	**LXA:** Lesson Extension Activity	**PU:** Power Up
	CA: Calculator Activity	**Inv:** Investigation	**PS:** Problem Solving	**TDA:** Test-Day Activity
	CT: Cumulative Test	**L:** Lesson	**PT:** Performance Task	**WP:** Written Practice
	ECE: End-of-Course Exam	**LS:** Learning Station		

Saxon Math Intermediate 4   © HMH Supplemental Publishers Inc.   **15**

COMMON CORE **Common Core State Standards**	*Saxon Math Intermediate 4* *Italic references indicate foundational.*
2. Compare two fractions with different numerators and different denominators, e.g., by creating common denominators or numerators, or by comparing to a benchmark fraction such as $\frac{1}{2}$. Recognize that comparisons are valid only when the two fractions refer to the same whole. Record the results of comparisons with symbols >, =, or <, and justify the conclusions, e.g., by using a visual fraction model.	**Lessons:** L56, WP56, WP57, WP59, WP61, WP62, WP64, WP66, WP71, WP78, Inv9, WP100, L103, WP103–WP109, WP111, *L116*, WP116–WP119 **Other:** LS56, BT4, LS103, LXA10, LS116, CT23, ET10, ECE

Build fractions from unit fractions by applying and extending previous understandings of operations on whole numbers.

3. Understand a fraction $\frac{a}{b}$ with a > 1 as a sum of fractions $\frac{1}{b}$.	
a. Understand addition and subtraction of fractions as joining and separating parts referring to the same whole.	**Lessons:** Inv9, L107, WP107, WP109, WP111–WP113, L114, WP114, WP116–WP118, L119, WP119, L120, WP120 **Other:** LS107, LS114, LS119, LS120, ECE
b. Decompose a fraction into a sum of fractions with the same denominator in more than one way, recording each decomposition by an equation. Justify decompositions, e.g., by using a visual fraction model.	**Lessons:** L89, WP89, WP90, Inv9, WP91–WP97, L104, WP104–WP110, WP112, WP116, WP118 **Other:** LS89, CT18, BT5, LS104, CT20, CT21
c. Add and subtract mixed numbers with like denominators, e.g., by replacing each mixed number with an equivalent fraction, and/or by using properties of operations and the relationship between addition and subtraction.	**Lessons:** *L89*, Inv9, L107, WP107–WP113, L114, WP114–WP117 **Other:** LS107, LXA9, LS114, CT23, ET9
d. Solve word problems involving addition and subtraction of fractions referring to the same whole and having like denominators, e.g., by using visual fraction models and equations to represent the problem.	**Lessons:** Inv9, L107, WP107–WP113, L114, WP114–WP117 **Other:** LS107
4. Apply and extend previous understandings of multiplication to multiply a fraction by a whole number.	
a. Understand a fraction $\frac{a}{b}$ as a multiple of $\frac{1}{b}$.	**Lessons:** *Inv9* **Other:** LXA6
b. Understand a multiple of $\frac{a}{b}$ as a multiple of $\frac{1}{b}$, and use this understanding to multiply a fraction by a whole number.	**Power Up:** PU106–PU112, PU116 **Lessons:** *L70, WP70, Inv9, L95, WP95, WP97–WP99, WP101–WP105, WP107–WP109, WP111, WP115–WP117, WP120* **Other:** BT4, *LS95*, LXA6, CT20, CT22
c. Solve word problems involving multiplication of a fraction by a whole number, e.g., by using visual fraction models and equations to represent the problem.	**Power Up:** PS116 **Lessons:** L70, WP70–WP73, WP81, WP83–WP86, WP89–WP94, L95, WP95–WP112, WP115–WP117, WP119, WP120 **Other:** LS70, LS95, LXA6, ET6, CT21

Understand decimal notation for fractions, and compare decimal fractions.

5. Express a fraction with denominator 10 as an equivalent fraction with denominator 100, and use this technique to add two fractions with respective denominators 10 and 100.[3]	**Lessons:** Inv4, *L50*, Inv9 **Other:** LXA2, ET2

[3]Students who can generate equivalent fractions can develop strategies for adding fractions with unlike denominators in general. But addition and subtraction with unlike denominators in general is not a requirement at this grade.

16 © HMH Supplemental Publishers Inc. *Saxon Math Intermediate 4*

COMMON CORE **Common Core State Standards**	*Saxon Math Intermediate 4* *Italic references indicate foundational.*
6. Use decimal notation for fractions with denominators 10 or 100.	**Lessons:** Inv4, WP41–WP43, WP47, WP51–WP56, WP61, WP63, WP64, WP67, L69, WP69, WP70, WP86, WP87, Inv9, WP91–WP95, WP97, WP98, WP100, L102, WP102–WP105, WP107, WP111, WP112, WP115, WP116, WP118 **Other:** CT21, CT22
7. Compare two decimals to hundredths by reasoning about their size. Recognize that comparisons are valid only when the two decimals refer to the same whole. Record the results of comparisons with the symbols >, =, or <, and justify the conclusions, e.g., by using a visual model.	**Lessons:** Inv4, WP41–WP43, WP47, Inv5, Inv9, L91, WP91, WP100, WP102 **Other:** LS91, BT7

Measurement and Data 4.MD

Solve problems involving measurement and conversion of measurements from a larger unit to a smaller unit.

1. Know relative sizes of measurement units within one system of units including km, m, cm; kg, g; lb, oz.; l, ml; hr, min, sec. Within a single system of measurement, express measurements in a larger unit in terms of a smaller unit. Record measurement equivalents in a two-column table.	**Power Up:** PU22–PU25, PU28–PU31, PU35, PU36, PU39–PU46, PU50, PU55, PU58, PU70, PU74, PU75, PU77, PU78, PU81, PU86–PU89, PU93, PU105, PU114 **Lessons:** Inv2, WP21–WP31, L32, WP32–WP35, WP37–WP39, L40, WP40–WP46, WP48–WP54, WP56–WP68, L69, WP69–WP76, L77, WP77–WP90, WP92, WP95, WP97, WP100, WP101, L102, WP102–WP105, WP108, WP110, WP112–WP114, WP116–WP120 **Other:** CT5, CT6, LS40, BT2, CT8, CT14, LS77, PT8, BT4, CT16, CT17, LS102, CT20, CT21, CT22, ECE
2. Use the four operations to solve word problems involving distances, intervals of time, liquid volumes, masses of objects, and money, including problems involving simple fractions or decimals, and problems that require expressing measurements given in a larger unit in terms of a smaller unit. Represent measurement quantities using diagrams such as number line diagrams that feature a measurement scale.	**Power Up:** PS42, PS51, PS56, PS67, PS73, PS78, PS87 **Lessons:** L19, Inv2, WP21–WP35, WP37–WP39, L40, WP40–WP43, WP45, WP46, WP48–WP54, WP56–WP68, L69, WP69, WP70, WP72, WP74–WP76, L77, WP77–WP82, L83, WP83–WP90, WP92, WP95, WP97, WP100, WP101, L102, WP102–WP105, WP108, WP110, Inv11, WP112–WP114, WP116–WP120 **Other:** PT1, LS19, TDA2, LXA1, TDA4, ET1, PT5, TDA5, PT7, PT12
3. Apply the area and perimeter formulas for rectangles in real world and mathematical problems.	**Power Up:** PU23, PU32, PU33, PU39, PU57, PU62, PU68 **Lessons:** Inv2, L21, WP21, WP28, WP30, Inv3, WP33, WP37, WP38, WP41, WP43, WP45, WP46, WP48, WP54, L55, WP55, WP56, WP59, L62, WP62, L69, WP72, WP74, WP77, WP83, WP89, WP97, WP103, WP104, WP106, L108, WP108, WP110, WP113, WP118 **Other:** BT4, CT18, LXA8, ET8, ECE

Represent and interpret data.

4. Make a line plot to display a data set of measurements in fractions of a unit ($\frac{1}{2}$, $\frac{1}{4}$, $\frac{1}{8}$). Solve problems involving addition and subtraction of fractions by using information presented in line plots.	**Other:** LXA7, ET7

Key:

BT: Benchmark Test	**ET:** Extension Test	**LXA:** Lesson Extension Activity	**PU:** Power Up
CA: Calculator Activity	**Inv:** Investigation	**PS:** Problem Solving	**TDA:** Test-Day Activity
CT: Cumulative Test	**L:** Lesson	**PT:** Performance Task	**WP:** Written Practice
ECE: End-of-Course Exam	**LS:** Learning Station		

Saxon Math Intermediate 4   © HMH Supplemental Publishers Inc.   **17**

COMMON CORE **Common Core State Standards**	*Saxon Math Intermediate 4* *Italic references indicate foundational.*

Geometric measurement: understand concepts of angle and measure angles.

5. Recognize angles as geometric shapes that are formed wherever two rays share a common endpoint, and understand concepts of angle measurement:	
a. An angle is measured with reference to a circle with its center at the common endpoint of the rays, by considering the fraction of the circular arc between the points where the two rays intersect the circle. An angle that turns through $\frac{1}{360}$ of a circle is called a "one-degree angle," and can be used to measure angles.	**Lessons:** *L78,* L81, WP84, WP85, WP87, WP94, WP97 **Other:** LS81
b. An angle that turns through *n* one-degree angles is said to have an angle measure of *n* degrees.	**Lessons:** *L78,* L81, WP84, WP85, WP87, WP94, WP97 **Other:** LS81, CT17, ECE
6. Measure angles in whole-number degrees using a protractor. Sketch angles of specified measure.	**Lessons:** *L81* **Other:** LXA4, LS81, ET4
7. Recognize angle measure as additive. When an angle is decomposed into non-overlapping parts, the angle measure of the whole is the sum of the angle measures of the parts. Solve addition and subtraction problems to find unknown angles on a diagram in real world and mathematical problems, e.g., by using an equation with a symbol for the unknown angle measure.	**Lessons:** L81 **Other:** LXA5, ET5

Geometry 4.G

Draw and identify lines and angles, and classify shapes by properties of their lines and angles.

1. Draw points, lines, line segments, rays, angles (right, acute, obtuse), and perpendicular and parallel lines. Identify these in two-dimensional figures.	**Lessons:** L23, WP23–WP36, WP38–WP43, L45, WP46, WP47, WP49, WP51, WP52, WP56–WP63, WP65, WP69, WP70, WP72, WP75, L78, WP84, WP85, WP87–WP89, L92, WP99, WP103, WP119, WP120 **Other:** LS23, CT5, BT3, LS78, CT16, TDA8, CT17, LS92, CT18, BT5, PT11, ECE
2. Classify two-dimensional figures based on the presence or absence of parallel or perpendicular lines, or the presence or absence of angles of a specified size. Recognize right triangles as a category, and identify right triangles.	**Power Up:** PS27 **Lessons:** L23, L45, WP45–WP50, WP52, WP55–WP61, WP65, L66, WP69–WP74, L78, WP78–WP80, WP82, WP83, WP85, WP87, WP88, WP90, WP91, L92, WP101, WP113, WP119, WP120 **Other:** LS78, CT16, LS92, TDA9, PT10, TDA10, ECE
3. Recognize a line of symmetry for a two-dimensional figure as a line across the figure such that the figure can be folded along the line into matching parts. Identify line-symmetric figures and draw lines of symmetry.	**Lessons:** L79, WP79, WP80, WP82, WP83, WP85, WP89, WP91, L92, WP94, WP95, WP100, WP105, WP109, WP112, WP113, WP116 **Other:** LS79, BT5

Key:

BT: Benchmark Test	**ET:** Extension Test	**LXA:** Lesson Extension Activity	**PU:** Power Up
CA: Calculator Activity	**Inv:** Investigation	**PS:** Problem Solving	**TDA:** Test-Day Activity
CT: Cumulative Test	**L:** Lesson	**PT:** Performance Task	**WP:** Written Practice
ECE: End-of-Course Exam	**LS:** Learning Station		

18   © HMH Supplemental Publishers Inc.   *Saxon Math Intermediate 4*

Saxon Math Intermediate 4
Extension Activity 1

• Using Diagrams to Solve Problems (CC.4.MD.2)

At the end of Lesson 40 complete the following activity.

Materials needed:

- *Activity Master 1*

We can use a diagram to help solve a problem.

> Destiny has $1\frac{1}{2}$ quarts of punch. She plans to use 1 cup of punch for each serving. How many one-cup servings can Destiny make?

Step 1: Use a number line diagram. Label the quarts.

Step 2: Use tick marks to show there are 4 cups in a quart.

Step 3: Draw and label the jumps on the diagram.

Step 4: Add to find the total number of one-cup servings Destiny can make.

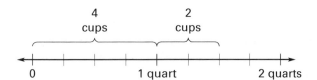

1 quart = 4 cups

$\frac{1}{2}$ quart = 2 cups

4 cups + 2 cups = 6 cups

So, Destiny can make _____ one-cup servings of punch.

Complete *Activity Master 1*.

Saxon Math Intermediate 4 © HMH Supplemental Publishers Inc.

Name _____

Activity Master 1

For use with Lesson 40 Extension Activity

• **Using Diagrams to Solve Problems**

Draw a number line diagram to help solve each problem.

1. Zeina put 4 liters of water in her aquarium. Then she added 500 milliliters more. How many milliliters of water did Zeina put in the aquarium altogether?

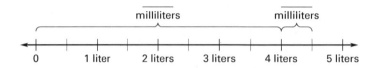

2. A muffin recipe uses $2\frac{1}{2}$ cups milk. Kendall's measuring cup only measures ounces. She knows 8 ounces is 1 cup. How many ounces of milk will Kendall use?

Saxon Math Intermediate 4

Saxon Math Intermediate 4
Extension Activity 2

INVESTIGATION 4

• Adding Fractions with Denominators 10 and 100

(CC.4.NF.5)

At the end of Investigation 4 complete the following activity.

Materials needed:

• *Activity Master 2*

Write an equal fraction with a denominator of 100.

1. $\frac{3}{10} = $ _____   2. $\frac{6}{10} = $ _____   3. $\frac{4}{10} = $ _____

We can use what we know to add fractions with denominators of 10 and 100.

Add: $\frac{4}{10} + \frac{7}{100}$

Step 1: Write an equal fraction with a denominator of 100. $\frac{4}{10} = \frac{40}{100}$

Step 2: Add the numerators.

The denominators stay the same. $\frac{40}{100} + \frac{7}{100} = \frac{47}{100}$

Shade the square to show the addition:

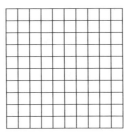

So, $\frac{4}{10} + \frac{7}{100} = $ _____

4. Add: $\frac{2}{10} + \frac{6}{100}$

Complete *Activity Master 2.*

Name _____

Activity Master **2**

*For use with **Investigation 4 Extension Activity***

• Adding Fractions with Denominators 10 and 100

Write an equal fraction with a denominator of 100.

1. $\dfrac{9}{10} =$ _____

2. $\dfrac{2}{10} =$ _____

3. $\dfrac{8}{10} =$ _____

Add:

4. $\dfrac{5}{10} + \dfrac{6}{100} =$ _____

5. $\dfrac{3}{10} + \dfrac{9}{100} =$ _____

© HMH Supplemental Publishers Inc.

Saxon Math Intermediate 4

Saxon Math Intermediate 4
Extension Activity 3

• Solving Comparison Problems (CC.4.OA.1, CC.4.OA.2, and CC.4.OA.3)

At the end of Lesson 49 complete the following activity.

Materials needed:

- *Activity Master 3*

In Lesson 49 we practiced solving word problems about equal groups. Look at Lesson 49, Example 1 and describe two different ways the problem can be solved.

In this activity we will practice solving word problems about comparisons. Read the problem below:

Emery folded 6 paper cranes. Harper folded 3 times as many paper cranes as Emery. How many paper cranes did Harper fold?

Underline the comparison statement in the problem. We can use a model to show the comparison.

Emery: | 6 cranes |

Harper: | 6 cranes | 6 cranes | 6 cranes |

3 times as many

The model compares the number of cranes Emery folded to the number of cranes Harper folded. It shows Harper folded **3 times as many** paper cranes as Emery.

We can write a multiplication equation to help solve the problem.

$3 \times 6 = n$ We translate the equation as **3 times 6 paper cranes = the number of paper cranes Harper folded.**

$n = 18$

So, Harper folded _____ paper cranes.

Complete *Activity Master 3*.

Name _____

Activity Master 3

For use with Lesson 49 Extension Activity

• Solving Comparison Problems

Draw a model and write an equation to solve each problem.

1. Spencer's dog weighs five times as much as Ben's cat. Ben's cat weighs 12 pounds. How much does Spencer's dog weigh?

2. Riker rode his bike 14 miles this week. That was twice the distance that he rode his bike last week. How many miles did Riker ride his bike last week?

© HMH Supplemental Publishers Inc.

Saxon Math Intermediate 4

Saxon Math Intermediate 4
Extension Activity 4

• Measuring and Drawing Angles (CC.4.MD.6)

At the end of Lesson 81 complete the following activity.

Materials needed:

- protractor
- *Activity Master 4*

Measure an angle using a protractor.

Step 1: Place the center point of the protractor on the vertex of the angle.

Step 2: Align the bottom ray of the angle with the 0° mark.

Step 3: Find where the other ray passes through the scale. Since the bottom ray passes through the 0° mark of the outer scale, the angle measure is read on that scale.

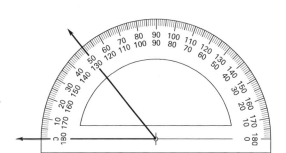

So, the angle measures 50°.

Follow the steps to measure the angle below:

_____ °

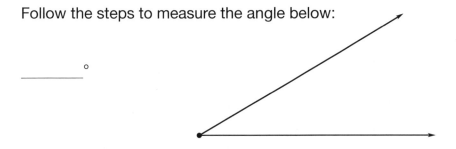

Draw an angle using a protractor.

Step 1: Use the straight edge of the protractor to draw a horizontal ray.

Step 2: Place the center point of the protractor on the endpoint. This will be the vertex of the angle. Align the ray with the 0° mark of either scale on the protractor.

Step 3: Using the same scale as the 0° mark, mark a point at 110°.

Step 4: Use a ruler to draw the other ray of the angle.

Complete *Activity Master 4*.

Name _____

Activity Master 4

For use with **Lesson 81 Extension Activity**

• Measuring and Drawing Angles

Use a protractor to find each angle measure.

1. _____°

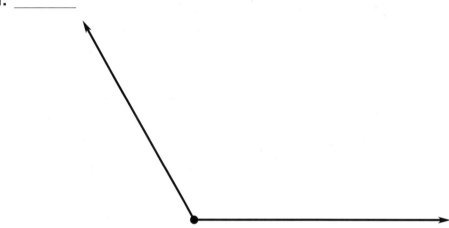

2. _____°

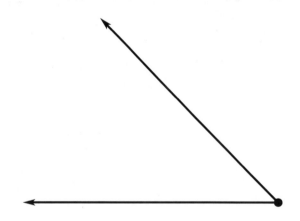

Use a protractor to draw each angle.

3. 75°

4. 105°

© HMH Supplemental Publishers Inc.

Saxon Math Intermediate 4

Saxon Math Intermediate 4
Extension Activity 5

• Joining and Separating Angles (CC.4.MD.7)

At the end of Lesson 92 complete the following activity.

Materials needed:

- *Activity Master 5*

Adjacent angles are side-by-side angles and share a common vertex and a common ray. We can use the measure of one adjacent angle and the measure of the larger angle formed by the outer rays of the adjacent angles to find the measure of the other adjacent angle.

We can use equations to find missing angle measures.

Find the unknown angle measure.

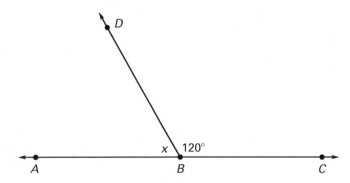

The diagram shows adjacent angles, ∠ABD and ∠CBD.

- The larger angle formed by the outer rays of the two adjacent angles is ∠ABC. It is a straight angle, so it measures _____°.
- ∠ABD and ∠CBD together form ∠ABC.
- m∠ABD + m∠CBD = m∠_____   We read m∠ABD as **measure of angle ABD**.
   x   +   120° = 180°   Think __?__ + 120 = 180, so 180 − 120 = __?__.
              x = 60°

So, m∠ABD = _____°.

Complete *Activity Master 5*.

Name _____

Activity Master 5

*For use with **Lesson 92 Extension Activity***

• Joining and Separating Angles

Write an equation to find the missing angle measure.

1. Two adjacent angles form a right angle. The measure of one angle is 35°. What is the measure of the other angle?

2. Two adjacent angles form a straight angle. One angle measures 78°. What is the measure of the other adjacent angle?

3. Two adjacent angles measure 135° and 45°. What type of angle do the two adjacent angles form?

4. Mr. Wong is using square tiles to make a design. He cuts each square tile into two equal pieces. The first piece cut has a 45° angle. What is the angle measure of the piece left over? Explain how you know.

© HMH Supplemental Publishers Inc.

***Saxon Math** Intermediate 4*

Saxon Math Intermediate 4
Extension Activity 6

• Multiplying a Fraction by a Whole Number

(CC.4.NF.4a, CC.4.NF.4b, CC.4.NF.4c)

At the end of Lesson 104 complete the following activity.

Materials needed:

- *Activity Master 6*

Read the problem below.

> Fran had 3 one-yard pieces of fabric. She used $\frac{3}{4}$ yard of each piece. How many yards of fabric did Fran use?

We can draw a picture to model this situation.

Step 1: Draw 3 rectangles to represent the 3 whole pieces of fabric.

Step 2: Draw lines horizontally to divide the 3 wholes into fourths.

Step 3: Shade $\frac{3}{4}$ of each rectangle. Count the number of shaded parts.

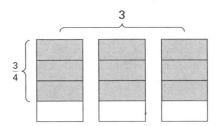

The model shows that 3 times $\frac{3}{4}$ is the same as 9 times $\frac{1}{4}$.

$3 \times \frac{3}{4} = 9 \times \frac{1}{4} = \frac{9}{4}$ or _____

So, Fran used _____ yards of fabric.

Complete *Activity Master 6*.

Name _____

For use with **Lesson 104 Extension Activity**

• **Multiplying a Fraction by a Whole Number**

Solve each problem. Draw a picture and write an equation to help solve the problem.

1. A recipe for cookies uses $\frac{2}{3}$ cup of butter. Maria plans to make 4 batches of cookies using the recipe. How many cups of butter will Maria need?

2. Sammy runs a distance of $\frac{4}{5}$ mile each weekday. How many miles does Sammy run each week?

© HMH Supplemental Publishers Inc.   *Saxon Math* Intermediate 4

Saxon Math Intermediate 4
Extension Activity 7

• Making a Line Plot to Display a Data Set (CC.4.MD.4)

At the end of Lesson 107 complete the following activity.

Materials needed:

- *Activity Master 7*

Vocabulary:

- **line plot:** A method of visually displaying a distribution of data values where each data value is shown as a dot or mark above a number line.

Read the problem below and make a line plot to display the data.

> The following distances were recorded (to the nearest eighth mile) to show how far members of the Woodlands Hiking Club hiked during one event.
>
> $\frac{7}{8}$ $\frac{4}{8}$ $\frac{4}{8}$ $\frac{6}{8}$ $\frac{3}{8}$ $\frac{5}{8}$ $\frac{5}{8}$ $\frac{7}{8}$ $\frac{4}{8}$ $\frac{5}{8}$ $\frac{6}{8}$ $\frac{7}{8}$ $\frac{5}{8}$

Step 1: Draw a number line. If we begin our number line at 0 and end it at 1 and divide it into fractional parts showing eighths, all of the data points can be included.

Step 2: Place an X on the number line for each data point. Since there are three data points that have a value $\frac{7}{8}$, we stack three Xs above $\frac{7}{8}$ on the number line.

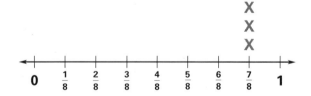

Distance Hiked
(to the nearest eighth mile)

Complete the line plot. Then refer to the line plot to answer problems 1–3. Remember to include labels with your answers.

1. Which distance was recorded the most often? _____

2. How many members hiked more than $\frac{4}{8}$ mile? _____

3. Write an equation and solve it to show the difference between the least distance and the greatest distance hiked. _____

Complete *Activity Master 7*.

Saxon Math Intermediate 4

Name _____

Activity Master 7

For use with **Lesson 107 Extension Activity**

• Making a Line Plot to Display a Data Set

Read the problem below.

Each member of Julia's science class planted a seed in a paper cup. After two weeks, the students measured and recorded the height of their plants to the nearest fourth inch. The heights were:

$5\frac{2}{4}$ $5\frac{3}{4}$ $6\frac{3}{4}$ $6\frac{1}{4}$ $5\frac{2}{4}$ $5\frac{3}{4}$ $6\frac{1}{4}$ $6\frac{2}{4}$ $5\frac{1}{4}$ $5\frac{3}{4}$ $6\frac{1}{4}$ $5\frac{3}{4}$

Make a line plot to display the data.

Plant Heights
(to the nearest fourth inch)

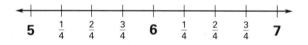

Refer to the line plot to answer problems 1–3.

1. Which height was recorded most often?

   _____

2. How many plants were over the height of 6 inches?

   _____

3. Write an equation and solve it to show the difference between the height of the shortest plant and the height of the tallest plant.

   _____

© HMH Supplemental Publishers Inc.

Saxon Math Intermediate 4

Saxon Math Intermediate 4
Extension Activity 8

• Finding Unknown Measures (CC.4.MD.3)

At the end of Lesson 108 complete the following activity.

Materials needed:

- *Activity Master 8*

Solve to find a missing measure when the area and the length or width are given.

> Mr. Harvey is building an outdoor deck that has a rectangular shape. The deck covers an area of 120 square feet. The length of the deck is 10 feet. What is the width of the deck?

Step 1: Use the formula for area of a rectangle. $A = l \times w$

Step 2: Fill in the measures you know. You know the
area is 120 square feet and the length is 10 feet. $120 = 10 \times w$

Step 3: Use a division sentence to find the missing
measure. Explain why. $120 \div 10 = w$

So, the width of the deck is _____ feet.

Solve to find a missing measure when the perimeter and the length or width are given.

> The perimeter of a rectangle is 20 centimeters. The length is 6 centimeters. What is the width of the rectangle?

Step 1: Use the formula for perimeter of a rectangle. $P = 2l + 2w$

Step 2: Fill in the measures you know. $20 = (2 \times 6) + (2 \times w)$

Step 3: Work backward to solve the problem.

Subtract the measure of the two lengths from
the perimeter. The difference is the sum of the
remaining two sides.   20 cm – 12 cm = _____ cm

Divide by 2 to find the width of one of the missing sides.   8 cm ÷ 2 = _____ cm

So, the width of the rectangle is _____ centimeters.

Complete *Activity Master 8*.

Saxon Math Intermediate 4   © HMH Supplemental Publishers Inc.   33

Name _____

Activity Master **8**

*For use with **Lesson 108 Extension Activity***

• Finding Unknown Measures

Use a formula to find the missing measure.

1. Robin is planting grass in a rectangular shaped space in her yard.
   The space has an area of 60 square feet and a width of 4 feet. What is
   the length of the space?

2. A park has a rectangular shaped playground. The perimeter of the
   playground is 280 feet. The length of the playground is 90 feet.
   What is the width of the playground?

© HMH Supplemental Publishers Inc.

Saxon Math Intermediate 4

LESSON 114

Saxon Math Intermediate 4
Extension Activity 9

• Adding and Subtracting Mixed Numbers (CC.4.NF.3c)

At the end of Lesson 114 complete the following activity.

Materials needed:

- fraction manipulatives
- *Activity Master 9*

In Lesson 114 we added mixed numbers by first adding the fraction parts and then adding the whole-number parts. This activity will show another method for adding and subtracting mixed numbers.

Use the steps below to solve this problem.

$2\frac{1}{3} + 3\frac{2}{3} =$

Step 1: Write each mixed number as an improper fraction.

Step 2: Add or subtract the improper fractions.

Step 3: Change improper fractions to a whole or mixed number.

You may use fraction manipulatives or draw a picture to model mixed numbers and improper fractions.

Step 1: $2\frac{1}{3} = \frac{7}{3}$

$3\frac{2}{3} = \frac{11}{3}$

Step 2: $\frac{7}{3} + \frac{11}{3} = \frac{18}{3}$

Step 3: $\frac{18}{3} =$ _____

Complete *Activity Master 9*.

Name _____

Activity Master 9

*For use with **Lesson 114 Extension Activity***

• Adding and Subtracting Mixed Numbers

Find each sum or difference by first writing each mixed number as an improper fraction. You may use fraction manipulatives or draw a picture to model mixed numbers and improper fractions.

1. $1\frac{4}{5} + 2\frac{3}{5} =$ _____

2. $3\frac{1}{6} - 1\frac{5}{6} =$ _____

3. Jeremy had a board that was $6\frac{1}{4}$ feet long. He cut off a piece that measured $4\frac{3}{4}$ feet. How long is the board that is left?

4. The painter used $2\frac{2}{3}$ gallons of paint on Monday. She used $1\frac{2}{3}$ gallons on Tuesday. How many gallons of paint did the painter use those two days?

© HMH Supplemental Publishers Inc.

***Saxon Math** Intermediate 4*

Saxon Math Intermediate 4
Extension Activity 10

• Comparing Fractions Using Common Denominators

(CC.4.NF.2)

At the end of Lesson 116 complete the following activity.

Materials needed:

- *Activity Master 10*

Fractions that do not have the same denominators may be difficult to compare. In Lesson 116 we used common denominators to rename fractions whose denominators are not the same. In this activity we will compare two fractions with different denominators using common denominators.

Look at Lesson 116, Example 1. Review the steps used to rename $\frac{2}{3}$ and $\frac{3}{4}$ using a common denominator.

Step 1: Write the fractions using a common denominator. $\frac{2}{3} = \frac{8}{12}$ and $\frac{3}{4} = \frac{9}{12}$

Step 2: Then write >, <, or = to compare the two fractions. $\frac{8}{12} \bigcirc \frac{9}{12}$ so $\frac{2}{3} \bigcirc \frac{3}{4}$

Use a fraction model to verify the comparison. When using models to compare fractions, the models that represent the whole must be the same size.

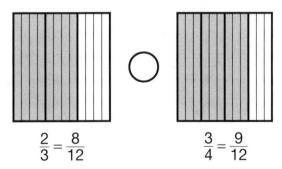

$\frac{2}{3} = \frac{8}{12}$ $\frac{3}{4} = \frac{9}{12}$

Complete *Activity Master 10*.

Saxon Math Intermediate 4   © HMH Supplemental Publishers Inc.   37

Name _____

Activity Master 10

*For use with **Lesson 116 Extension Activity***

• Comparing Fractions Using Common Denominators

Rename each pair of fractions using a common denominator. Then write >, <, or = to complete the comparison. Draw a fraction model to verify the comparison.

1. $\frac{1}{3}$ ◯ $\frac{1}{4}$

2. $\frac{2}{5}$ ◯ $\frac{1}{2}$

© HMH Supplemental Publishers Inc.

Saxon Math Intermediate 4

Name _____

Score _____

Extension Test **1**

*For use with **Cumulative Test 8***

• Using Diagrams to Solve Problems

Draw a number line diagram to help solve each problem. Fill in the circle with the correct answer.

1. Mr. Collins bought 2 gallons of motor oil. He used 5 quarts of the oil to replace the oil in his car. How many quarts of motor oil does Mr. Collins have left?

Ⓐ 5 quarts

Ⓑ 3 quarts

Ⓒ 2 quarts

Ⓓ 1 quart

2. Container A holds 1 liter. Container B holds 200 milliliters. How many B containers would it take to fill Container A?

Ⓐ 5

Ⓑ 4

Ⓒ 3

Ⓓ 2

3. Leigh is mixing paint. She combined 3 quarts of blue paint with 1 pint of red paint to make purple paint. How many pints of purple paint does Leigh have?

Ⓐ 13 pints

Ⓑ 10 pints

Ⓒ 7 pints

Ⓓ 5 pints

4. Kevin bought a pint of milk. He used 4 ounces of milk for a recipe. How many ounces of milk does Kevin have left?

Ⓐ 16 ounces

Ⓑ 12 ounces

Ⓒ 8 ounces

Ⓓ 4 ounces

Saxon Math Intermediate 4

© HMH Supplemental Publishers Inc.

Name _____

Score _____

Extension Test **2**

For use with **Cumulative Test 9**

• Adding Fractions with Denominators 10 and 100

Fill in the circle with the correct answer.

Which shows an equal fraction?

1. $\dfrac{9}{10} =$ _____

   Ⓐ $\dfrac{1}{100}$

   Ⓑ $\dfrac{9}{100}$

   Ⓒ $\dfrac{80}{100}$

   Ⓓ $\dfrac{90}{100}$

2. $\dfrac{30}{100} =$ _____

   Ⓐ $\dfrac{1}{10}$

   Ⓑ $\dfrac{2}{10}$

   Ⓒ $\dfrac{3}{10}$

   Ⓓ $\dfrac{4}{10}$

Add:

3. $\dfrac{2}{10} + \dfrac{3}{100} =$ _____

   Ⓐ $\dfrac{5}{100}$

   Ⓑ $\dfrac{23}{100}$

   Ⓒ $\dfrac{5}{10}$

   Ⓓ $\dfrac{23}{10}$

4. $\dfrac{7}{10} + \dfrac{2}{100} =$ _____

   Ⓐ $\dfrac{9}{10}$

   Ⓑ $\dfrac{72}{10}$

   Ⓒ $\dfrac{72}{100}$

   Ⓓ $\dfrac{9}{100}$

© HMH Supplemental Publishers Inc.

Saxon Math Intermediate 4

Name _____

Score _____

Extension Test 3

*For use with **Cumulative Test 10***

• Solving Comparison Problems

Draw a model and write an equation to solve each problem. Fill in the circle with the correct answer.

1. Mr. Smith is three times as tall as his daughter. His daughter is 2 feet tall. How tall is Mr. Smith?

Ⓐ 4 feet

Ⓑ 5 feet

Ⓒ 6 feet

Ⓓ 7 feet

2. Pamela has 42 rocks in her collection. Ethan has 4 times as many rocks in his collection. How many rocks does Ethan have in his collection?

Ⓐ   46 rocks

Ⓑ   88 rocks

Ⓒ 126 rocks

Ⓓ 168 rocks

3. A watch costs $30. A camera costs five times as much as the watch. How much does the camera cost?

Ⓐ $35

Ⓑ $80

Ⓒ $120

Ⓓ $150

4. The first bus had 40 passengers. That was twice as many passengers as the second bus had. How many passengers did the second bus have?

Ⓐ 80 passengers

Ⓑ 42 passengers

Ⓒ 20 passengers

Ⓓ   2 passengers

Saxon Math Intermediate 4

© HMH Supplemental Publishers Inc.

Name _____

Score _____

Extension Test 4

For use with **Cumulative Test 16**

• **Measuring and Drawing Angles**

Use a protractor to find each angle measure. Fill in the circle with the correct answer.

1. What is the measure of the angle?

   Ⓐ 140°
   Ⓑ 130°
   Ⓒ 50°
   Ⓓ 40°

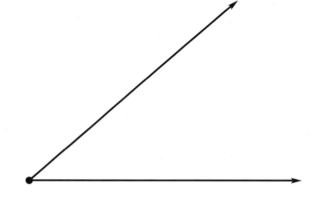

2. What is the measure of the angle?

   Ⓐ 100°
   Ⓑ 95°
   Ⓒ 80°
   Ⓓ 75°

© HMH Supplemental Publishers Inc.

Saxon Math Intermediate 4

Name _____

Score _____

Extension Test **5**

*For use with **Cumulative Test 19***

• Joining and Separating Angles

Write an equation to find the missing angle measure. Fill in the circle with the correct answer.

1. Two adjacent angles form a right angle. The measure of one angle is 30°. What is the measure of the other angle?

Ⓐ 150°

Ⓑ 70°

Ⓒ 60°

Ⓓ 50°

2. Two adjacent angles form a straight angle. One angle measures 126°. What is the measure of the other adjacent angle?

Ⓐ 66°

Ⓑ 54°

Ⓒ 44°

Ⓓ 36°

3. Two adjacent angles measure 25° and 65°. What type of angle do the two adjacent angles form?

Ⓐ straight

Ⓑ right

Ⓒ acute

Ⓓ obtuse

4. A diagram on Micah's test shows a straight angle separated into two adjacent angles. Micah measures one angle and finds the angle measurement is 105°. What type of angle is the other adjacent angle?

Ⓐ straight

Ⓑ right

Ⓒ acute

Ⓓ obtuse

Saxon Math Intermediate 4   © HMH Supplemental Publishers Inc.

Name _____

Score _____

Extension Test 6

*For use with **Cumulative Test 20***

• Multiplying a Fraction by a Whole Number

Draw a picture and write an equation to help solve each problem. Fill in the circle with the correct answer.

1. Cara wants to buy 9 new books. Her mother said she can only buy $\frac{2}{3}$ of that number. How many new books can Cara buy?

    Ⓐ 9 books

    Ⓑ 6 books

    Ⓒ 3 books

    Ⓓ 2 books

2. The corner deli uses $\frac{1}{2}$ pound each of 5 different deli meats to make a sandwich tray. How many pounds of meat altogether does the deli use for a sandwich tray?

    Ⓐ $1\frac{1}{2}$ pounds

    Ⓑ 2 pounds

    Ⓒ $2\frac{1}{5}$ pounds

    Ⓓ $2\frac{1}{2}$ pounds

3. Kim bought 3 one-gallon containers of different flavors of yogurt. She used $\frac{1}{2}$ of each gallon to make smoothies. How many gallons of yogurt did Kim use for the smoothies?

    Ⓐ $2\frac{1}{2}$ gallons

    Ⓑ 2 gallons

    Ⓒ $1\frac{1}{2}$ gallons

    Ⓓ $1\frac{1}{3}$ gallons

4. Daniel read 10 books last month. Two-fifths of the books were fiction. How many fiction books did Daniel read last month?

    Ⓐ 4 books

    Ⓑ 5 books

    Ⓒ 6 books

    Ⓓ 8 books

© HMH Supplemental Publishers Inc.

***Saxon Math** Intermediate 4*

Name _____

Score _____

Extension Test 7

For use with **Cumulative Test 21**

• **Making A Line Plot to Display a Data Set**

Refer to the line plot to answer the questions below. Fill in the circle with the correct answer.

The students in Liam's class spread the fingers of their left hand as far apart as possible and measured the distance from the tip of the little finger to the tip of the thumb. The lengths, measured to the nearest eighth inch, are shown on this line plot:

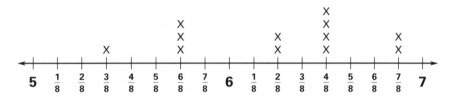

Hand Lengths
(to nearest eighth inch)

1. How many students measured their hands?

   Ⓐ 8
   Ⓑ 12
   Ⓒ 20
   Ⓓ 28

2. Which length appears most often?

   Ⓐ $5\frac{3}{8}$ inches
   Ⓑ $5\frac{6}{8}$ inches
   Ⓒ $6\frac{4}{8}$ inches
   Ⓓ $6\frac{7}{8}$ inches

3. How many students' hands measured less than 6 inches?

   Ⓐ 4
   Ⓑ 5
   Ⓒ 7
   Ⓓ 8

4. The *range* is the difference between the greatest value and the smallest value plotted. What is the range of the data shown?

   Ⓐ $6\frac{7}{8}$ inches
   Ⓑ $5\frac{3}{8}$ inches
   Ⓒ $2\frac{4}{8}$ inches
   Ⓓ $1\frac{4}{8}$ inches

Saxon Math Intermediate 4   © HMH Supplemental Publishers Inc.

Name _____

Score _____

Extension Test 8

For use with **Cumulative Test 22**

• Finding Unknown Measures

Use a formula to find the missing measure. Fill in the circle with the correct answer.

1. Roger drew a rectangle that has an area of 88 square centimeters. The length of the rectangle is 8 centimeters. What is the width of the rectangle Roger drew?

   Ⓐ 36 centimeters
   Ⓑ 19 centimeters
   Ⓒ 11 centimeters
   Ⓓ 10 centimeters

2. Patty plays volleyball on a rectangular court. The perimeter of the court measures 60 yards. The width of the court is 10 yards. What is the length of the volleyball court?

   Ⓐ 40 yards
   Ⓑ 30 yards
   Ⓒ 20 yards
   Ⓓ 15 yards

3. An oil painting has an area of 132 square inches. The length of the painting is 12 inches. What is the width of the painting?

   Ⓐ 11 inches
   Ⓑ 22 inches
   Ⓒ 23 inches
   Ⓓ 46 inches

4. A rectangular shaped swimming pool has a perimeter of 46 meters. The length of the pool is 14 meters. What is the width of the pool?

   Ⓐ 32 meters
   Ⓑ 26 meters
   Ⓒ 18 meters
   Ⓓ 9 meters

© HMH Supplemental Publishers Inc.

Saxon Math Intermediate 4

Name _____

Score _____

Extension Test   **9**

*For use with **Cumulative Test 23***

• Adding and Subtracting Mixed Numbers

Find each sum or difference by first writing each mixed number as an improper fraction. Fill in the circle with the correct answer.

1. $2\frac{3}{8} + 1\frac{7}{8} =$

  Ⓐ $3\frac{1}{4}$

  Ⓑ $4\frac{1}{4}$

  Ⓒ $4\frac{3}{8}$

  Ⓓ $5\frac{1}{8}$

3. Marsha spent $1\frac{1}{4}$ hours practicing the piano one day and $1\frac{3}{4}$ hours practicing on another day. Altogether, how many hours did Marsha practice the piano?

  Ⓐ   2   hours

  Ⓑ $2\frac{3}{4}$ hours

  Ⓒ   3   hours

  Ⓓ $3\frac{1}{4}$ hours

2. $6\frac{1}{4} - 3\frac{3}{4} =$

  Ⓐ $3\frac{1}{2}$

  Ⓑ $3\frac{1}{4}$

  Ⓒ $2\frac{3}{4}$

  Ⓓ $2\frac{1}{2}$

4. The park has two walking paths. The length of the East Path is $3\frac{5}{8}$ miles. The length of the West Path is $2\frac{2}{8}$ miles. How much longer is the East Path?

  Ⓐ $1\frac{3}{8}$ miles

  Ⓑ $1\frac{5}{8}$ miles

  Ⓒ   4   miles

  Ⓓ $5\frac{7}{8}$ miles

***Saxon Math** Intermediate 4*

© HMH Supplemental Publishers Inc.

Name _____

Score _____

Extension Test **10**

For use with **Cumulative Test 23**

• Comparing Fractions Using Common Denominators

Rename each pair of fractions using a common denominator. Then choose a symbol to complete the comparison.

1. $\frac{3}{8}$ ◯ $\frac{1}{2}$

   Ⓐ >

   Ⓑ <

   Ⓒ =

3. $\frac{5}{6}$ ◯ $\frac{3}{4}$

   Ⓐ >

   Ⓑ <

   Ⓒ =

2. $\frac{1}{2}$ ◯ $\frac{5}{6}$

   Ⓐ >

   Ⓑ <

   Ⓒ =

4. $\frac{1}{2}$ ◯ $\frac{5}{10}$

   Ⓐ >

   Ⓑ <

   Ⓒ =

© HMH Supplemental Publishers Inc.

Saxon Math Intermediate 4

Lesson Extension Activity Answers

Lesson Extension Activity 1

6

Lesson Extension Activity 2

1. $\frac{30}{100}$ 2. $\frac{60}{100}$ 3. $\frac{40}{100}$

See student work. $\frac{47}{100}$

4. $\frac{2}{10} = \frac{20}{100}$; $\frac{20}{100} + \frac{6}{100} = \frac{26}{100}$

Lesson Extension Activity 3

18

Lesson Extension Activity 4

30°; See student work.

Lesson Extension Activity 5

180°; m∠ABC; 60°

Lesson Extension Activity 6

$2\frac{1}{4}$; $2\frac{1}{4}$

Lesson Extension Activity 7

See student work.

1. $\frac{5}{8}$ mile

2. 9 members

3. $\frac{7}{8} - \frac{3}{8} = \frac{4}{8}$ or $\frac{1}{2}$ mile

Lesson Extension Activity 8

12; 4

Lesson Extension Activity 9

6

Lesson Extension Activity 10

<; <; <

Activity Master Answers

Activity Master 1

1. See student work.
   4000; 500; 4500 milliliters

2. See student work. 20 ounces

Activity Master 2

1. $\frac{90}{100}$ 2. $\frac{20}{100}$ 3. $\frac{80}{100}$

4. $\frac{5}{10} = \frac{50}{100}$; $\frac{50}{100} + \frac{6}{100} = \frac{56}{100}$

5. $\frac{3}{10} = \frac{30}{100}$; $\frac{30}{100} + \frac{9}{100} = \frac{39}{100}$

Activity Master 3

1. See student work. 60 pounds

2. See student work. 7 miles

Activity Master 4

1. 120°

2. 45°

3. See student work.

4. See student work.

Activity Master 5

1. $90 - 35 = 55$; 55°

2. $180 - 78 = 102$; 102°

3. $135 + 45 = 180$; straight angle

4. 45°; Answers will vary; Possible answer: Square tiles have corners that are 90°. If the tiles are cut so that the corner is 45°, the measure of the other corner is $90° - 45° = 45°$.

Activity Master 6

1. See student work.
   $4 \times \frac{2}{3} = 8 \times \frac{1}{3} = \frac{8}{3}$ or $2\frac{2}{3}$ cups

2. See student work.
   $5 \times \frac{4}{5} = 20 \times \frac{1}{5} = \frac{20}{5}$ or 4 miles

Activity Master 7

See student work.

1. $5\frac{3}{4}$ inches   2. 5 plants

3. $6\frac{3}{4} - 5\frac{1}{4} = 1\frac{2}{4}$ or $1\frac{1}{2}$ inches

Activity Master 8

Possible steps for Problems 1 and 2:

1. $A = l \times w$
   $60 = l \times 4$
   $60 \div 4 = l$
   $l = 15$ feet

2. $P = 2l + 2w$
   $280 = (2 \times 90) + (2 \times w)$
   $280 - 180 = 100$
   100 feet $\div 2 = 50$ feet

Activity Master 9

1. $4\frac{2}{5}$ 2. $1\frac{2}{6}$ or $1\frac{1}{3}$

3. $1\frac{2}{4}$ or $1\frac{1}{2}$ feet   4. $4\frac{1}{3}$ gallons

Activity Master 10

1. $\frac{1}{3} > \frac{1}{4}$; See student work.

2. $\frac{2}{5} < \frac{1}{2}$; See student work.

Saxon Math Intermediate 4   © HMH Supplemental Publishers Inc.

Extension Test Answers

Extension Test 1

1. Ⓑ 3 quarts
2. Ⓐ 5
3. Ⓒ 7 pints
4. Ⓑ 12 ounces

Extension Test 2

1. Ⓓ $\frac{90}{100}$
2. Ⓒ $\frac{3}{10}$
3. Ⓑ $\frac{23}{100}$
4. Ⓒ $\frac{72}{100}$

Extension Test 3

1. Ⓒ 6 feet
2. Ⓓ 168 rocks
3. Ⓓ $150
4. Ⓒ 20 passengers

Extension Test 4

1. Ⓓ 40°
2. Ⓐ 100°

Extension Test 5

See student work.

1. Ⓒ 60°; 90 − 30 = 60
2. Ⓑ 54°; 180 − 126 = 54
3. Ⓑ right; 25 + 65 = 90
4. Ⓒ acute; 180 − 105 = 75

Extension Test 6

1. Ⓑ 6 books
2. Ⓓ $2\frac{1}{2}$ pounds
3. Ⓒ $1\frac{1}{2}$ gallons
4. Ⓐ 4 books

Extension Test 7

1. Ⓑ 12
2. Ⓒ $6\frac{4}{8}$ inches
3. Ⓐ 4
4. Ⓓ $1\frac{4}{8}$ inches

Extension Test 8

1. Ⓒ 11 centimeters
2. Ⓒ 20 yards
3. Ⓐ 11 inches
4. Ⓓ 9 meters

Extension Test 9

1. Ⓑ $4\frac{1}{4}$
2. Ⓓ $2\frac{1}{2}$
3. Ⓒ 3 hours
4. Ⓐ $1\frac{3}{8}$ miles

Extension Test 10

1. Ⓑ <
2. Ⓑ <
3. Ⓐ >
4. Ⓒ =

Saxon Math Intermediate 4